Images and Reflections

Images and Reflections

Dennis van Westerborg

Temeron Books Inc.

First published in 1991 by
Temeron Books Incorporated
Box G 399
Calgary, Alberta
T3A 2G3
(403) 283-0900
Fax (403) 283-6947

Canadian Cataloguing in Publication Data
van Westerborg, Dennis, 1930 -
Images and reflections

Includes index.
ISBN 1-895510-00-7

1. van Westerborg, Dennis, 1930- I. Title.
ND249.V36A4 1991 759.11 C91-091572-5

Art Consultants: *Linda Carney and Allan Jensen*
Editor: *Theresa Goulet*
Designers: *Clayton Warholm and Theresa Goulet*

Printed in Canada by Paperworks Press Limited. Images and Reflections is printed on Quintessence Dull Text 80#, with a Bayside Linen cover, Quintessence Gloss 100# dust jacket, and Gainsborough end sheets. The typeface is Zapf Chancery.

Dedication

To Natalie Munro

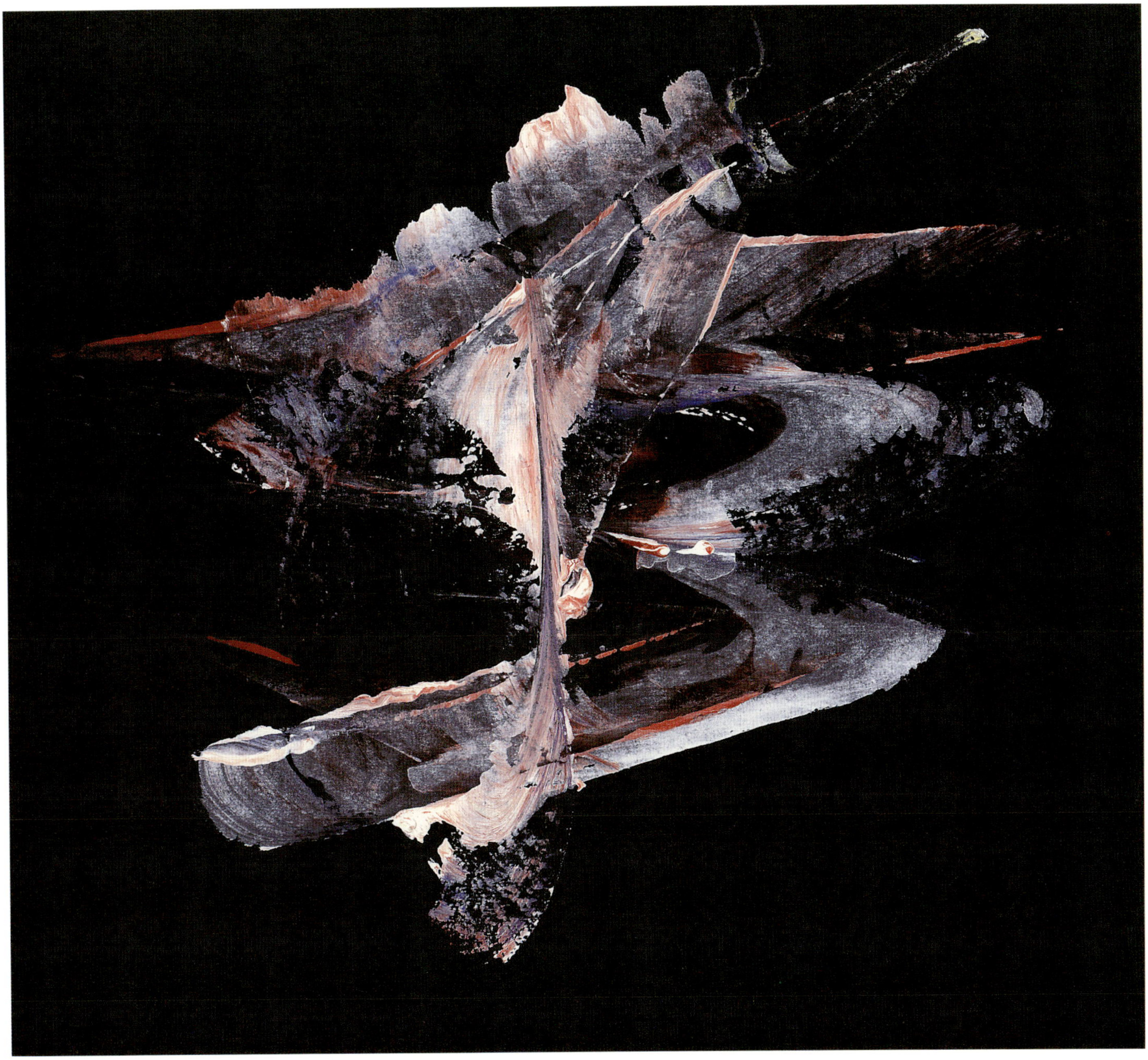

"Dennis van Westerborg looks tremendously gifted, but also strangely uneven in view of his giftedness."

Clement Greenberg
New York, September 14, 1991

Introduction

The artist's business consists neither in copying nature nor in rendering life, but in faithfully representing himself. - Dennis van Westerborg

Dennis van Westerborg's work is particularly notable for the range of styles and media it encompasses. His artwork includes watercolors, inks, oils, acrylics and collages - from delicate calligraphic pieces to bold abstracts. In this book our challenge was to bring these diverse works together into a cohesive whole. As you look through the book, you will notice that although each image leads into the next through a common color, line, or shape, we have deliberately avoided placing similar pieces together. Instead, we have arranged the artwork so that turning the pages will be an adventure for the reader.

Although text and artwork have been carefully juxtaposed, neither is meant to "illustrate" the other. We believe that each poem, aphorism, and artwork can stand on its own merit.

Theresa Goulet, Editor/Designer

What I Know About God

A juggler who keeps Himself busy
balancing nature;
Judging from all the lakes and oceans on earth,
He too likes playing in puddles;
Ever since making men and women so different,
He has figured as the world's foremost sexist;
Having begun as an unknown dramatist, He
worked his way up to noted producer and
director of history;
He is known by the kind of universe He keeps;
Mother Nature is His wife.
Welcome to God's amusement park, the world.

You need not hate humanity to qualify as a cynic.
Just stand back and watch in amusement.

Other people offer excuses for their conduct,
whereas we give explanations for ours.

Let's skip the ceremonial courtesies and
proceed directly to the hostilities.

Women are said to feel rather than think.
Men sometimes think and sometimes feel and
sometimes they do neither.

Thought generally doesn't acknowledge its debt
to feeling until we try to think straight
while feeling depressed.

Thoughts furnish the concepts, events the context,
and feelings the accompaniment to
the dance of life.

I would rather think new thoughts than see new sights.

One of the things new under the sun is the opportunity to look at things in a new light and throw the perceptions into fresh combinations.

What good is perfect eyesight when you are inwardly blind?

Spring Cleaning

I do my spring cleaning
in the fall,
the fall cleaning
doesn't get done
at all.

War: crime raised to a higher power

Nations hate one another - not for sharing the same feelings but for expressing them in different languages.

International convention governs the treatment of prisoners of war, yet stops short of preventing the wars that make prisoners.

What would become of war if neutrality broke out all over the world?

Now that our bodies have been introduced,
are our minds also going to meet?

Just one attraction is enough to bring the sexes
together, whereupon a hundred forces
set to work to pull them apart.

Given time and patience, any passing fancy may
eventually ripen into lasting infatuation.

As soon as you come to expect it,
it's no longer romance.

In religion, as in business, it pays to bypass the distributor and deal directly with the manufacturer.

God is always on the side of those who don't know whose side He's on.

Atheism is better than no religion at all.

Thrifty worshippers everywhere prefer generic to brand name religion.

Stained glass: windows that prevent you from seeing God in His true light.

Kinds of Music

Performed by wind and sea,
birds, insects, frogs, wolves;
spoken as well as sung or played,
heard at times in a child's laugh
and then in the casual words
of a voice well loved,
or again in the lilt and inflection
of unknown tongues,
the music of many kinds.

Music is the poetry of sound, dancing the poetry of movement, painting the poetry of color.

I think of music as sound in color,
pastel shades as colors in a minor key,
and poetry as sculpture in thought.

Until poets take hold of it, a language is just a medium of exchange. Until musicians shape sound, it is only disturbed air.

It was shockingly irresponsible of God to invent life without leaving user instructions or program notes.

If we didn't exist, do you think God would have to invent us?

The management of the universe accepts no responsibility for misplaced expectations, injured feelings, or lost illusions.

Life has no meaning - you have to bring your own.

Whether God plays dice with the cosmos or not, it's certainly His to play dice with.

*Governments must take care how
they deal with poets: metaphors are
sometimes mightier than armies.*

*It's no use killing a man for his ideas -
they'll only come back in another mind.*

Revolution: ideas in heat

*Some streets are easy to find unless
directions are given; some concepts are
perfectly clear until they are defined.*

Mind Over Matter

Not mere things,
but thoughts on things
and thoughts on thoughts
delight me.

It takes science many years of exhaustive study to reach the conclusion that it would have been better not to have disturbed Nature in the first place.

When the last pesticide will ultimately have failed of effect, the insects shall inherit the earth, the meek having nothing to do with it.

The earth comes rent-free, complete with running water, central heating and air conditioning. All the landlord asks in return is that we keep the place clean.

For me despoiling the earth constitutes the real and unwritten "sin", and the day when our planet finally ceases to be habitable because of human mismanagement is the real Judgment Day.

Women should not lower themselves to seek equality with men, who, after all, are to blame for creating the mess the world is in.

Nature's plan includes even those who do not perceive her plan.

Fur coats look much more interesting when there is an animal inside, preferably wild and free.

Although solitary nature hikes are most enjoyable, this can be corrected by instituting organizational meetings and bylaws.

At the Grave of Time

For the sake of old times I laid
a wreath today
on an unmarked grave where
I used to sit by the hour
and run time through my hands like sand
and watch it die.

A thing becomes really funny only after someone tells you not to laugh because it's not funny.

Unintended humor and unintended love are the best kind.

Nothing illuminates a topic better than levity; nothing improves nonsense more than treating it in dead earnest.

Laughing at myself would be a lot easier if I were someone else.

Since my mood has changed,
reality isn't what it used to be.

I sit on the edge of reality and dangle my legs in dreams. Or is it the other way round?

If God didn't exist we wouldn't be here
to question His existence.

Reality is the ultimate illusion.

Matter has a mind of its own.

Memorials

One man leaves descendants and wealth,
yet his name is forgotten;
another leaves nothing - except
the legacy of his mind,
and is remembered.

Perspiration begets inspiration
as quantity begets quality.

Sometimes I get big ideas but
by and large I think small.

Just because I get my best ideas in the middle
of the night doesn't mean they're good.

The only way to find inspiration
is to stop looking for it.

"Mind for sale. Has never been used."

Everywhere wildlife is being destroyed by industrialization and everywhere in the corporate offices of industrialists hang splendid paintings to celebrate wildlife. If the substance is being sacrificed, at least the image is intact.

Wildness is retreating, barbarism is advancing.

The big question now is no longer economic or political or even socio-economic, but whether the earth will still be habitable in a hundred years.

The "will of the people" is not so much a collective sentiment as a fiction created by speech writers, public relations firms, and advertising agencies. Thus democratic elections are little more than a contest between propagandists. They are the highpriests of democracy.

Sudden religious "conversions" merely exchange one passion for another. Sudden social "emancipations" merely exchange one form of bondage for another.

Every majority began as a minority; every excess began as reform.

Irrevocables

A tree cut down,
A well run dry,
A word once spoken,
A love gone by.

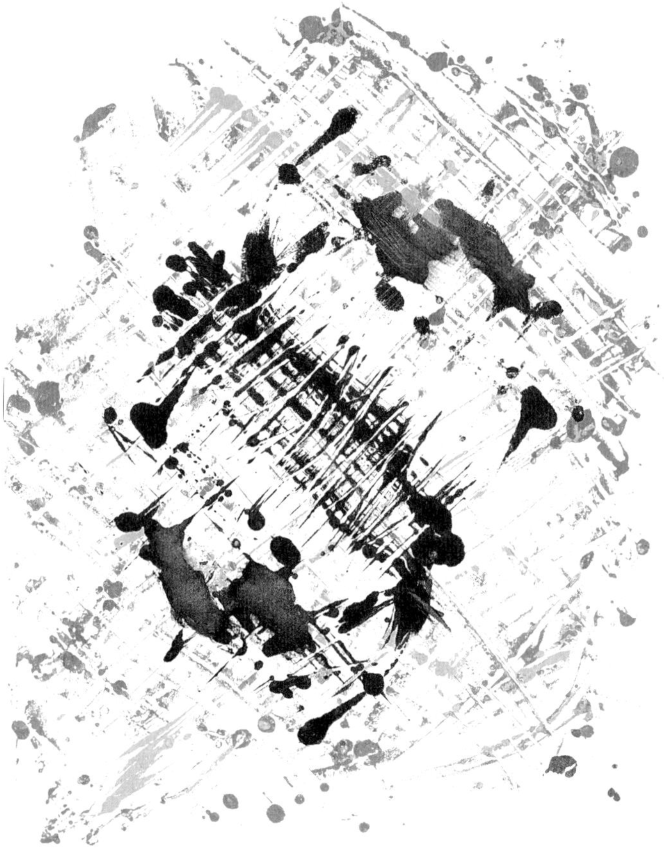

Art imitates art.

If he's driven to produce, the artist need take no responsibility for the quality of his work. After all, he couldn't help it.

It saddens me to think that at this very moment someone somewhere is producing beautiful art which I may never see.

If I were perfect I wouldn't be an artist.

You can take a man to court;
bringing him to justice is something else.

The right to plead not guilty when you know you are
engenders not justice but dishonesty.

The law appears often more concerned with
technicalities, correct procedure and prescribed
form than with the pursuit of justice.

There may indeed be elected government, but
democracy is an illusion; laws and judges are
concrete enough, but justice is an abstraction.

You were not able to capture Lincoln's spirit,
so you set up his bust in your library.

Being a good listener is not necessarily of benefit
unless there is talk worth hearing.

There are always people to talk with
but not often those you can talk to.
Prepositions do make a difference.

There is something to be said
for leaving something unsaid.

The Conference

The birds held an early-morning conference;
Robin called the meeting to order,
Old Crow read the minutes.
Sparrows then formed a committee to
settle a housing dispute, whereupon
a discussion ensued on the merits of
everything in general.
Magpies kept raising procedural objections,
but spring was thought to be a good thing
all round and was unanimously voted in
for another year.
There being no further business, the
meeting adjourned at dawn.

It's the little things in life that trip me up;
the big ones I take in stride.

Adversity betokens Providence no less than prosperity, for in either case you mattered enough for the gods to have bothered with you and thereby demonstrated their interest in your continuing education.

Without the negative, the positive has no meaning.
Without evil, good goes unrecognized.

Fear death? Not on your life!

Not until the spirit leaves my body shall I finally be able to rise above my surroundings.

If medical science succeeds in finding a cure for cancer, we shall have no choice but to die of other causes instead.

There has to be another existence after this, if for no other reason than to satisfy all the longings introduced but left unsatisfied in this life.

*If there were but one language and one religion,
nations would have to find more compelling
reasons for feuding.*

*The wars fought to rectify national borders
are sometimes sparked by nothing more
than love of symmetry.*

*Prescription for perpetual peace: ensure that
leaders who declare war fight it as well.*

*War pretends to defend civilization but
it does so by barbarous means.*

Overlooked

I studied botany but never noticed
trees talking to each other or
bushes conferring in the breeze;
I professed religion but never worshipped
in Nature's shrine;
I demanded miracles and saw them
every day.

Being born against their wishes tends to make people hard to get along with.

Until you've lived your whole life, it's hard to tell whether you were wellborn.

Surviving childhood can be quite worthwhile, especially if one grows up too.

When you are all choked up listening to Brahms,
what does music theory matter?

When you say it with music
you needn't say anything else.

Melancholy music is the sweetest -
to melancholy minds.

What kind of dinner music goes with chicken?

Does nihilism mean nothing to you?

Once you put your faith in scepticism
you need have no further doubts.

To deny the existence of mind serves only
to affirm that you have one.

Boredom is like nothing on earth.

Blood runs thicker than water,
but money runs thicker than blood.

Getting rich is easier than living rich.

By cheating me out of character-building hardships,
wealth has impoverished me.

Have you ever wondered where "old money"
originally came from?

If you must live long, try not to get old.

*After men get too old to set a bad example,
their greatest pleasure is to persuade
young men to do likewise.*

*At one time I had a weak mind but a strong back.
Now both are weak.*

The Long Way Round

I sought wisdom in the schools,
but found only knowledge;
I sought God in churches,
but found only dogma and ritual;
I sought worship,
and found it in nature;
I sought peace,
and found it in myself.

Creative people have little control over their lives, but fortunately they rarely pose public danger.

Creative persons are often amoral, or immoral, or egotistical, or eccentric, or neurotic, or impotent, or homosexual, etc., whereas those who have nothing glaringly the matter with them frequently don't matter much in the long run.

If great poets are proverbially tinged with madness, then what does this say for the sanity of those who enjoy reading their work?

Novelty is the soul of romance,
routine the death of it.

For man and woman to become one flesh
is easy enough; learning to be of one mind
takes a little longer.

As intervals between elections test the quality
of government, so intervals between sexual
intercourse test conjugal life.

But for the "refractory period" of the human male
after sexual intercourse, life would be but one long
copulation, and the gods would be denied the
entertainment we now provide.

Although the practice of medicine need not be life-threatening, doctors are doing their best.

When it comes to choosing the illness that's right for you, opt for a little of everything rather than too much of any one.

I owe my good health to all the medicine I haven't taken.

Pursuit of happiness is making me sad; obsession with health is making me ill.

Transudation, or the indestructibility of matter: you eat garlic one day and meet it again the next, coming through your armpits.

Music

Conspiracy of rhythm
And chemistry of tone:
Primordial appeal
And language all its own.

Donate your body to science: pathologists will take more interest in you than your family ever did.

If you ever feel rejected, consider the case of Father of Pearl.

Loved or hated - either is better than getting no attention at all.

I don't necessarily have to look up to you
because you are taller than I.

Of all your qualities I envy
your lack of envy the most.

I don't care whether you're fair-minded
so long as you agree with me.

If I have to keep reminding you - forget it!

Labels

Nature, Energy, or Chance,
arbitrarily defined, are
but different names for God,
forms of universal mind.

Cows have been sacred long enough -
it's now cats' turn again.

Dogs can be ordered to heel, but have you ever
heard of cat obedience classes?

Dogs have masters, cats have slaves.

Cats always get their way, but they manipulate us
with such perfect grace that we don't even know
we've been had.

When we fall in love it is generally with an illusion. Occasionally, however, we may come to love the person who inspired it.

Would the girl of my dreams at least have the decency to come forward and identify herself!

Lingerie by itself is not very exciting - there's nothing in it.

What I lack in mystique I try
to make up in technique.

I'm not conceited enough to think I'm humble.

I need to slow down but I won't rush into it.

Having got the formalities out of the way,
we can now get on with the trivialities.

We live by feelings but survive by thought.

It never hurts to feel good.

When I think I'm thinking I only manipulate my feelings; when I feel, I get to the heart of the matter.

Feelings alone are never ridiculous unless we try to defend them with reason.

Depth of feeling alone gives no more guarantee of truth than heat alone guarantees light.

In your search for mind-altering substances,
have you tried thought?

Learning would be such fun
if only I didn't have to do it!

Repetition is the stepmother of
learning, learning, learning.

No lesson was intended -
that's why I learned so much.

Learning from others' mistakes has taken
all the fun out of making my own.

Carriers

A thought that had slipped from some mind
into print, unnoticed,
lay there embalmed for a long time,
pressed between the pages,
but then it accidentally fell
out one year, got taken up and
revived by another mind.

The main difference between us and the ancients is not intelligence, but technology. "Western civilization", for the most part, means only mechanization.

Progress: I don't know what it is but I'm all for it.

Nothing heralds the imminent collapse of civilization like finding misspelled words in a dictionary.

Consumption: currently, the process of "using up"; formerly, one of "wasting away".

If I don't do what I want, who will?

*Whether the human will is free or not,
we're free to think we're free.*

*Without self-possession I soon become
the possession of others.*

*One measure of freedom is the number
of crutches you can live without.*

*Having fewer desires than other people
does not make me better, only freer.*

Good listening leaves me speechless.

The longer the explanation,
the shorter the attention span.

Talk was cheap, so I bought more
than I could afford.

Giving advice is always ill-advised.

Philosophical beliefs, however biased, are often
more convincing when baldly stated than
when supported by long arguments.

It may be consoling to read that great men have great faults. Less consoling, however, is the sudden discovery of great faults in oneself without the concomitant redeeming greatness.

Humility is nothing to be proud of.

As you can see, there's very little difference between man's importance and his impotence.

Bread of affliction: a quaint dietary supplement not unlike humble pie.

The Permanence of Change

The world seems like a river,
Life constantly flows through;
Its contents ever changing,
Its patterns never do.

Life constantly renews itself,
Takes never-ending pains;
We age, change, and depart:
Only the pattern remains.

The wisdom that's forward-looking and always up to date is found only by going back in time.

Wisdom can be discerned, garnered, enjoyed, expressed - but not passed on. To live by it is the most one can hope for.

Will I ever have enough training to know whether I have enough?

Wisdom exists not to be possessed but only for the love of it.

Described in sufficient detail, even boredom eventually becomes interesting.

Extreme moderation goes too far.

Total objectivity presupposes no presuppositions.

Complexity is nothing more than simplicity gone to seed.

If it's all the same I'd rather not have equality.

His claim to fame rests on having renounced fame.

After a Dangerous Illness

Feebly you step out of doors
as if out of prison
and savor as never before
the freedom, sunshine, trees
and sense as never before
or since
the keen exhilaration of
consciousness heightened,
of life made aware of itself.

In long marriages love often dies prematurely,
but fortunately habit and convenience
carry on as if nothing had happened.

Living together militates against loving together.

Love is too precious a thing
to spend itself on marriage.

One can't be too careful in the choice of enemies
or people one falls in love with.

Science has finally confirmed that a direct
relationship exists between alimony and acrimony.

I must get out of this life and into something more comfortable.

Life is preoccupied with new arrivals and departures. Persons in between are expected to fend for themselves.

Why make death more of a riddle than birth?

Perhaps the fear of death springs, after all, only from resentment at the lack of prior consultation.

Even if it doesn't lead anywhere life is still good experience to list on your resume.

It took me all my life to learn to know myself, but by then it was too late.

We got kicked out of Eden and consequently have been suffering withdrawal symptoms ever since.

"Current edition of humanity soon to be discontinued. Publisher not expected to re-issue."

To The Art Critic

Artistically perfect,
the shading and light,
The technique is correct,
The colors are right.

So you jury the sky
The sunset aglow,
With a critical eye
As if judging a show.

But untutored am I
In the technical part,
All I see in the sky
Is majestic God's art.

Arbitrariness has to start somewhere.

Consistency is essential only in things
that don't matter for long.

Allow me to make this impassioned plea
for objectivity.

Far from being forgotten, his name lives on
in the annals of oblivion.

Clearly the situation is not at all clear.

It's easier to attach new labels to old concepts than to re-examine the concepts themselves; it's easier to coin new catchwords than to think new thoughts.

When I run out of words I have to think.
When I run out of thought I use words.

Language belongs primarily to poets.
Others use it at their own risk.

For the author life is nothing but
a vast peg to hang his opinions on.

Evidently managers of symphonies everywhere hire regulars to cough professionally at concerts, particularly when the music is soft and low.

Musicology knows the theory, musicality has the pleasure.

When all else fails to express the inexpressible mood of the moment, music often comes in handy.

Talking is a waste of time when one could be listening - to music.

Opinions tend to be more interesting than mere facts, just as artistic impressions may be more interesting than mere snapshots.

Simply stated, the facts are too complex to be simply stated.

Facts can't be trusted: let them speak for themselves and they stab you in the back.

Better the fertility of speculation than the sterility of facts.

Modern conveniences have liberated us from meaningless drudgery by freeing up more time for leisure activities which are just as meaningless.

It's either too hot, too cold, too wet, or too dry. Why not abolish weather altogether?

Perhaps the flowers that have no scent appeal to insects that have no sense of smell.

A Whimsy

One day, when the numbers rose in rebellion
against their handlers, a violent explosion
ripped through the calculations, hurling
fragments of fractions and formulas
in all directions.
Several disciplines were seriously
disrupted,
creating a state of predictive uncertainty.
Dialogue is continuing.

Luck is infatuated with those
who don't believe in it.

Ignoring or defying the worst invites the worst;
expecting it tends to ward it off.

When you expect nothing the smallest boon
comes as a pleasant surprise.

It's amazing how much can be gained
when you have nothing to lose.

To an Atheist

Your sky is but an unsigned canvas, even though
the artist's signature flames in every sunset,
looms large in every storm,
authenticates the whole exhibition.

The quest for unity has divided us more than ever.

Under democracy you vote not for the greater good
but for the lesser evil.

I am solidly on the side of nonpartisanship.

When nationalism walks in the door
national sanity flies out the window.

Think twice before following your conscience.
It may be unconstitutional.

Living to a ripe old age affords plenty of time for preaching the brevity of life.

Young and foolish, old and foolish - it's just a matter of time.

It's true I'm getting older, but haven't I been all my life?

In his youth a man may be able to get by on looks alone; once he starts losing his hair he is on his own.

No better proof of steady growth than a double chin.

To forgive is one thing, but to forget amounts to throwing away valuable experience.

It costs dearly to see clearly.

Life has taught me everything I know - but I know very little.

I came across a stockpile of time somebody had saved. Evidently it had never been used.

Time goes by in no time. Where does it go when it runs out?

Take as much time as you want - there'll still be enough left for future generations.

Truth does not depend on newly discovered facts.
There will always be new facts,
but not necessarily new truths.

Truisms: shallow platitudes held by other people,
as distinct from my own deep truths.

Honest truth: "true" facts as opposed to false
Naked truth: facts that ought to be
ashamed of themselves

Beauty hides everywhere in full view,
truth lurks well hidden.

Patriotism

I stand to attention
for earth and sky,
and the wail of the wind
is my national anthem.

How much in reality ultimately turns into illusion!

If your dreams are sweet what more
can reality do for you?

Reality is just another word for
observer consciousness.

The answers to the major questions
are more questions.

Some questions contain their own answers.

Loneliness wants to break out of its prison,
solitude locks the door from inside.

A friend in need is a friend to avoid.

The chief value of other people is that they
give me something to think about.

Problem-solving is fun as long as it's not my own.

Trust an artist to choose beauty over duty or to jettison facts for the sake of a metaphor!

When you are moved by an image, touched by a note in the right key, is it not argument enough?

From an artistic perspective it matters not how well we live nor yet how long. How much of us survives is really all that counts in the long run.

Don't wear your nonconformity prematurely - wait till it's in fashion.

If it is true that no two people are alike and glad of it, then why does everyone long to be someone else?

Comparisons are proverbially odious, and yet how else can we take our own measure?

The Hunter as Hunted

He hunted for the sake of trophies,
she hunted him in turn, making
him her trophy,
domesticated, stuffed and mounted.

The contemplation of science ultimately leads into religion; the arts already dwell in the realm of the godlike.

Where recognition comes only posthumously, the only good artist is a dead one.

Beware of artists who will sacrifice reality for beauty. They are capable of anything!

Juxtaposition is everything: in art, in music, in poetry, in life generally.

Color Illustrations

All artworks are from the artist's collection and are untitled except as noted. Sizes shown are height x width.

Front dust jacket
watercolor
13 cm x 8 cm

Page

1 watercolor
15 cm x 14 cm

2 acrylic
11 cm x 8 cm

5 acrylic
33 cm x 25 cm

6 mixed media
14 cm x 12 cm

8 watercolor and ink
12 cm x 11 cm

11 monoprint
23 cm x 20 cm

13 oil
35 cm x 28 cm

15 oil
24 cm x 22 cm

17 collage,
mixed media
29 cm x 45 cm

19 monoprint
21 cm x 15 cm

21 oil
40 cm x 35 cm

23 monoprint
25 cm x 20 cm

25 mixed media
11 cm x 7 cm

27 "Flying Fish"
watercolour
15 cm x 15 cm

29 "Calligraphetus"
acrylic
36 cm x 30 cm

31 oil
14 cm x 9 cm

33 mixed media
21 cm x 15 cm

35 watercolor
15 cm x 10 cm

37 oil
28 cm x 46 cm

39 watercolor
12 cm x 21 cm

41 "Fractal Coastline"
monoprint
9 cm x 14 cm

43 watercolor and ink
21 cm x 15 cm

45 oil
42 cm x 31 cm

47 mixed media
14 cm x 10 cm

49 oil
23 cm x 31 cm

51 oil
18 cm x 24 cm

53 oil
28 cm x 21 cm

55 oil
40 cm x 27 cm

57 collage, monoprint
38 cm x 52 cm

59 oil
25 cm x 28 cm

61 watercolor and ink
16 cm x 10 cm

63 watercolor and ink
25 cm x 17 cm

65 watercolor
18 cm x 15 cm

67 collage,
mixed media
33 cm x 25 cm

69 monoprint
18 cm x 24 cm

71 mixed media
19 cm x 13 cm

73 monoprint
12 cm x 10 cm

75 monoprint
28 cm x 21 cm

77 watercolor and ink
18 cm x 21 cm

79 oil
43 cm x 35 cm

81 watercolor and ink
21 cm x 15 cm

83 "Centrifugal Force"
oil
28 cm x 39 cm

85 watercolor
7 cm x 10 cm

87 oil
21 cm x 15 cm

89 oil
13 cm x 8 cm

91 watercolor
15 cm x 10 cm

93 oil
32 cm x 25 cm

95 watercolor and ink
22 cm x 15 cm

97 oil
21 cm x 19 cm

99 watercolor and ink
24 cm x 19 cm

101 watercolor and ink
14 cm x 17 cm

103 monoprint
18 cm x 28 cm

105 watercolor and ink
12 cm x 12 cm

107 oil
33 cm x 26 cm

109 oil
40 cm x 25 cm

111 watercolor and ink
22 cm x 19 cm

113 watercolor
17 cm x 15 cm

115 monoprint
31 cm x 23 cm

117 oil
21 cm x 15 cm

119 watercolor and ink
12 cm x 11 cm

121 acrylic
14 cm x 10 cm

123 watercolor
15 cm x 15 cm

125 oil
35 cm x 28 cm

127 "Prairie Cactus"
watercolor
10 cm x 14 cm

129 oil
16 cm x 12 cm

131 oil
28 cm x 19 cm

133 acrylic and ink
38 cm x 33 cm

135 oil
14 cm x 25 cm

137 watercolor
7 cm x 9 cm

139 monoprint
35 cm x 18 cm

141 watercolor and ink
12 cm x 13 cm

143 oil
21 cm x 15 cm

145 oil
33 cm x 25 cm

147 watercolor and ink
26 cm x 18 cm

149 "Tsunami"
oil
7 cm x 11 cm

151 oil
15 cm x 21 cm

153 oil
11 cm x 14 cm

155 oil
27 cm x 18 cm

157 oil
36 cm x 28 cm

159 monoprint
26 cm x 21 cm

161 monoprint
23 cm x 18 cm

163 watercolor
11 cm x 7 cm

165 oil
11 cm x 7 cm

167 monoprint
22 cm x 15 cm

169 acrylic
33 cm x 25 cm

171 watercolor and ink
21 cm x 15 cm

173 oil
11 cm x 7 cm

175 oil
21 cm x 15 cm

177 watercolor
10 cm x 15 cm

179 oil
7 cm x 11 cm

181 "Pirouette"
watercolor and ink
19 cm x 11 cm

183 monoprint
28 cm x 21 cm

185 monoprint
21 cm x 15 cm

187 watercolor and ink
25 cm x 20 cm

189 acrylic
33 cm x 25 cm

190 oil
18 cm x 25 cm

Back dust jacket
oil
25 cm x 16 cm